Preservation of Food

Home Canning, Preserving, Jelly-making, Pickling, Drying

by

Olive Elliot Hayes

APPLEWOOD BOOKS
Bedford, Massachusetts

Preservation of Food

was originally published in

1919

ISBN: 978-1-4290-1053-5

Thank you for purchasing an Applewood book.
Applewood reprints America's lively classics—
books from the past that are still of interest
to the modern reader.
For a free copy of
a catalog of our
bestselling
books,
write
to us at:
Applewood Books
Box 365
Bedford, MA 01730
or visit us on the web at:
For cookbooks: foodsville.com
For our complete catalog: awb.com

Prepared for publishing by HP

PRESERVATION OF FOOD.

HOME CANNING.

WHAT is canning? Canning is the method of preservation based upon a simple plan of keeping bacteria away from food products, and has been devised in the last century. This has come more and more into common use, until to-day it is employed to an almost incredible extent. The method spoken of is canning. The food is not treated by any antiseptic for the prevention of bacterial growth, but reliance is placed simply upon devices for keeping all bacteria from it. If this can be done, the food will not be subject to their action and will never spoil.

Bacteria are almost universally distributed in earth, air, and water. This fact makes it extremely difficult to protect food from their action, and without special devices it is quite impossible to do so. All food material—meats, fruits, or vegetables—is sure to contain bacteria when it reaches the home or the canning-factory. From some source, either air, water, or earth, every kind of food material is sure to become contaminated. We must recognize, then, that bacteria will be found with absolute certainty in every kind of fresh food. Hence the process of keeping food by protecting it from bacteria must consist of two steps: (1) Some means must be devised for removing the bacteria already present in the food; (2) the access of all other bacteria must be absolutely prevented. If these two objects can be accomplished, the food will be protected from bacterial action and thus protected may be preserved indefinitely. No limit has ever been found, and we have no reason for questioning that it might be preserved for centuries without any subsequent change, provided it could be kept absolutely free from the attack of micro-organisms. This method therefore offers almost unlimited possibilities in the way of preserving food for future use. It demands care in its application, but the results when properly obtained are permanent.

Success in canning necessitates the destruction of these organisms. A temperature of 160° to 190° Fahr. will kill yeasts and moulds. Bacteria are destroyed at a temperature of 112° Fahr. held for the proper length of time. The destruction of these organisms by heat is called sterilization.

METHODS OF CANNING.

There are four principal methods of home canning. These are:—

(1.) Single-period or Cold-pack Method.

(2.) Fractional or Intermittent Sterilization Method.
(3.) Open-kettle or Hot-pack Method.
(4.) Cold-water Method.

Of these methods the one recommended for home use is the Single-period or Cold-pack Method. It is much the best because of its simplicity and effectiveness, and in this book detailed instructions are given for its use. The outlines of the various methods are as follows:—

(1.) *Single-period or Cold-pack Method.*—The prepared vegetables or fruits are blanched in boiling water or live steam, then quickly cold-dipped and packed at once into hot jars and sterilized in boiling water or by steam-pressure. The jars are then sealed, tested for leaks, and stored. Full details of this method are given in the following pages.

(2.) *Fractional or Intermittent Sterilization Method.*—Vegetables are more difficult to can than fruits because of the presence of spore-bearing bacteria, which are more resistant to heat than yeasts or moulds. These bacteria will live and decompose vegetables even with the exclusion of air. They reproduce by spores which retain vitality for a long time even at boiling temperature, and on cooling will germinate. For this reason, therefore, in order to completely sterilize some vegetables it is necessary to boil for one hour on three successive days. The boiling on the first day kills all the living bacteria, but does not kill the spores. As the jar cools the spores germinate and the boiling on the second day kills the fresh crop of bacteria. The third boiling is to ensure perfect sterilization. This method is known as the Intermittent Method and is strongly recommended for the canning of peas, beans, corn, asparagus, greens, pumpkin, and squash. Variations in soil, moisture, and climatic conditions from year to year make cause of failure one year when success has always attended the One-period or Cold-pack Method.

(3.) *Open-kettle or Hot-pack Method.*—Vegetables or fruits are cooked in an open kettle and packed in hot jars. There is always danger of spores and bacteria being introduced on spoons or other utensils while the jars are being filled. This method should never be used in canning vegetables; even with fruits it is not as desirable as cold-pack.

(4.) *Cold-water Method.*—Rhubarb, cranberries, gooseberries, and sour cherries because of their acidity are often canned by this method. The fruits are washed, put in sterilized jars, cold water is added to overflowing, and the jars are then sealed. This method is *not always successful,* as the acid content varies with the ripeness and the locality in which the fruits are grown.

HOME CANNING.

COLD-PACK METHOD—ONE PERIOD.

The One-period or Cold-pack Method is to be preferred to all other methods, as it decreases the work of canning and with an occasional exception is just as effective as the Intermittent Method. It is the method used in canning-factories where the food is canned under pressure, but it may be used in the home, and the following equipment used:—

Equipment.

(1.) A wash-boiler, new garbage-pail, or galvanized tub may be used. The cover should be tight fitting. To ensure a tight-fitting cover an inch rim should be soldered on the cover (as is shown in Fig. 1), so that it will give a tighter fit and make the boiler hold the

Fig. 1. A false bottom of wood (or of wire) placed in an ordinary wash-boiler makes a good outfit. Note rim on cover.

steam. Instead of a rim the cover may be adjusted with cloth (as in Fig. 2), and this will serve the same purpose.

(2.) A false bottom should be used which would allow a ¾-inch space at the bottom for the circulation of water. The rack prevents the jars from coming in contact with the hot metal of the bottom of the boiler; therefore preventing the breakage of the glass jar. The false bottom could be of wire netting of size suitable for the boiler used. A rack made of strips of wood similar to the one shown might be used, or a board which has been bored full of holes at regular intervals.

Fig. 2. Adjustment of ordinary cover with cloth gives a tight fit and boiler will hold the steam.

Terms Explained.

(1.) **Scalding** is pouring water over the food in order:—
(*a.*) To loosen the skin:
(*b.*) To eliminate objectionable acids and acrid flavour:

Fig. 3. Blanching with wire basket.

(*c.*) To start the flow of colouring-material.

(**2.**) **Blanching** is boiling the food in water or steaming. Use a cheese-cloth bag, or a wire frying-basket or strainer, to lower the food into the water. The blanching process is used:—

(*a.*) To loosen the skin:

(*b.*) To eliminate acids and acrid flavour:

(*c.*) To reduce bulk:

(*d.*) To make the Intermittent or Fractional Method unnecessary.

Fig. 4. After being blanched the product is immediately dipped in cold water.

(**3.**) **Cold-dip** is to chill quickly by dipping into cold water the fruits or vegetables. Purposes of cold-dip are:—

(*a.*) To harden the pulp under the skin and thus permit the removal of the skin:

(*b.*) To coagulate the colouring-matter and to make it harder to dissolve during the sterilization period:

(*c.*) To make it easier to handle the products in packing.

Jars and Lids must be Thoroughly Sterilized.

Jars which have not been thoroughly cleaned when they are emptied of their former contents, or which had contained mouldy fruit and had not been thoroughly washed and sterilized, contain a number of resting spores.

See that the jars are sound, without groove or nicks, and that they are perfectly clean. Thoroughly scald with boiling water. A good method is to wash the jars and put them in a large kettle of warm water. Let the water come to a boil, and boil for five minutes. Lift jars from kettle with a long sterilized stick as required.

See that rubbers and tops are sound and let them stand in scalding water for a few minutes. Always use new rubbers of a good thick quality. Adjust the rubbers before filling the jar.

DIRECTIONS FOR CANNING VEGETABLES BY THE COLD-PACK METHOD.

(1.) Choose vegetables that are young and have made a quick growth.

(2.) Do not use very dirty vegetables.

(3.) Can vegetables as soon as possible after picking. This is particularly necessary with asparagus, peas, beans, and corn.

(4.) Clean the vegetables and prepare them as for cooking.

(5.) Grade the vegetables if there is much variation in size, so the contents of each jar will be as nearly uniform in size as possible.

(6.) Do not attempt to handle too large a quantity of vegetables at once, especially in hot weather. The various steps in the canning process must be followed in rapid succession to prevent loss of flavour from what commercial canners know as "flat sour."

(7.) Blanch or scald the vegetables by plunging them into a large quantity of boiling water. (A wire basket or a cheese-cloth may be

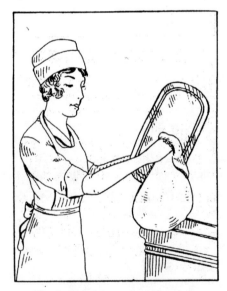

Fig. 5. Blanching with cheese-cloth.

HOME CANNING.

used for this, as is shown in cut.) The blanching or scalding should be continued just long enough to make the vegetables sufficiently flexible to pack easily or to loosen the skins sufficiently to allow them to be quickly scraped off. Spinach and certain other delicately flavoured greens should be blanched in steam, instead of in boiling water, until they are thoroughly shrunken. (Method for doing this is shown in Fig. 6.) One-half teaspoon soda may be added to each gallon of water to help set the colour of green vegetables.

(8.) Chill the outside of the vegetables by immersing them quickly in a large vessel of cold water. Do not attempt to cool the vegetables by this cold-dip.

(9.) Pack the vegetables firmly in sterilized, tested jars to within ½ inch of the top.

(10.) Add 1 teaspoon of salt to each quart jar and ½ teaspoon of salt to each pint jar. Some vegetables, especially corn, are improved by the addition of a small amount of sugar as well.

Fig. 7. Position of clamp during sterilization.   Fig. 8. Position of clamp after sterilization.

(11.) Fill the jars with boiling water to within ½ inch of the top.

(12.) Place a new rubber on each jar, adjust the cover, and partially seal it.

(13.) Place in sterilizer in which there is sufficient warm water to cover tops of jars about 1 inch. If this is done, little or no liquid is lost from jars during sterilization. Do not allow jars to touch in sterilizer, as this will usually cause breakage.

(14.) Sterilize the jars for the required length of time, counting from the time the water begins to boil.

(15.) Keep the water boiling during the sterilization period.

(16.) Vegetables may be successfully sterilized by the Cold-pack One-day Method. It is very occasional that vegetables grown in the Western Provinces contain bacteria which necessitates the Three-day or Intermittent Method. If, however, results are to be absolutely certain, the intermittent process of sterilization as described is safest unless the steam-pressure canner is used.

(17.) Remove the jars from the sterilizer, seal them, and invert them to cool, being careful to avoid a draught on the jars; but cool them as quickly as possible, especially in canning peas, beans, corn, asparagus, and greens.

(18.) When jars are cold wash them and set away.

Asparagus.

Wash, scrape off scales and tough skin. With a string bind together enough for one jar. Blanch tough ends from 5 to 10 minutes, then turn so that the entire bundle is blanched 5 minutes longer. Cold-dip. Remove string. Pack, with tip ends up. To each quart jar add 1 teaspoon salt, 2 tablespoons vinegar, and cover with boiling water. (The vinegar assists in retarding the growth of certain bacteria characteristic in asparagus.) Put on rubber and adjust top. Sterilize 120 minutes in hot-water bath. Remove, complete seal, and cool.

With steam-pressure outfit sterilize 60 minutes at 5 to 10 lb. pressure.

Cauliflower.

Wash and divide head into small pieces. Soak in salted water 1 hour, which will remove insects if any are present. Blanch 3 minutes. Cold-dip and pack in jars. Add 1 teaspoon salt to each quart jar and cover with boiling water. Put on rubber and adjust top. Sterilize 60 minutes in hot-water bath. Remove, complete seal, and cool.

With steam-pressure outfit sterilize 30 minutes at 5 to 10 lb. pressure.

Carrots.

Select small, tender carrots. Leave an inch or two of stems; wash; blanch 5 minutes and cold-dip. Then remove skin and stems. Pack whole or in slices. Add 1 teaspoon salt to each quart jar and cover with boiling water. Put on rubber and adjust top. Sterilize 90 minutes in hot-water bath. Remove, complete seal, and cool.

With steam-pressure outfit sterilize 60 minutes at 5 to 10 lb. pressure.

Cabbage and Brussels Sprouts.

The method is the same as for cauliflower, except that the vegetables are not soaked in salted water. Blanch 5 to 10 minutes. Sterilize 120 minutes in hot-water bath.

With steam-pressure outfit sterilize 60 minutes at 5 to 10 lb. pressure.

Beets.

Wash beets and twist off stalks, but do not cut or break root, as this causes the loss of colour. Boil until more than three-quarters cooked. Blanch in cold water and rub off skins and stems. They may be packed in jar whole or cut in halves or quarters. If the root is cut lengthwise rather than across, not so much of the colour and flavour is lost. Add 2 tablespoons salt, ¼ cup sugar, and ¼ cup vinegar to each quart jar. (Vinegar helps to retain the colour and retard the growth of bacteria.) Cover with boiling water. Put on rubber and adjust top. Sterilize 30 minutes in hot-water bath. Remove, complete seal, and cool.

With steam-pressure outfit sterilize 10 minutes at 5 lb. pressure.

Corn.

Use the corn when it is freshly picked. Remove the husks and silk, blanch tender ears 5 minutes, older ears 10 minutes. Cold-dip and cut from cob. Pack lightly to within ½ inch of the top of the jar, as corn swells during sterilization. Add 1 teaspoon salt, and 1 tablespoon sugar if desired. Cover with boiling water. Put on rubber and adjust top. Sterilize 180 minutes in hot-water bath. Remove, complete seal, and cool. When canning on the cob, pack jars alternating butts and tips. Sterilize corn on the cob 4 hours in hot-water bath. Canning corn on the cob, except for exhibition purposes, is a waste of space.

With steam-pressure outfit sterilize 90 minutes at 5 to 10 lb. pressure.

Greens.

Allow to stand in salty water for 10 minutes to free from insects. Wash in several waters until no dirt can be felt in the bottom of the pan. Blanch in steam 15 minutes (mineral matter is lost if blanched in water). Cold-dip. Cut in small pieces and pack, or pack whole. Do not pack too tightly. Add 1 teaspoon salt to each jar and cover with boiling water. Put on rubber and adjust top. Sterilize 120 minutes in hot-water bath. Remove, complete seal, and cool.

With steam-pressure outfit sterilize 60 minutes at 5 to 10 lb. pressure.

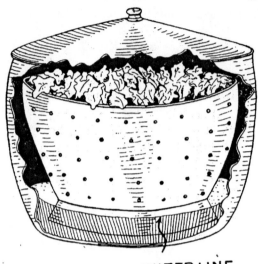

Fig. 6. Spinach and greens should not be blanched in hot water. They should be blanched in steam from 10 to 15 minutes. This cut shows a simple method of blanching in steam, by placing them in a colander in a kettle with tightly fitting cover. There should be not more than an inch or so of water on the bottom of the kettle and the water should not touch the greens. A steam-pressure canner is excellent for use in blanching greens.

Parsnips.

The method is the same as for carrots.

Peas.

Those which are not fully grown are best for canning. Shell, blanch 5 to 10 minutes, and cold-dip. Pack in jar. Add 1 teaspoon salt and cover with boiling water. If the jar is packed too full, some of the peas will break and give a cloudy appearance to the liquid. Put on rubber and adjust top. Sterilize 180 minutes in hot-water bath. Remove, complete seal, and cool.

With steam-pressure outfit sterilize 60 minutes at 5 to 10 lb. pressure.

Peppers.

Wash, stem, and remove seeds. Blanch 5 to 10 minutes. Cold-dip and pack in jars. Add 1 teaspoon salt, cover with boiling water.

Put on rubber and adjust top. Sterilize 120 minutes in hot-water bath. Remove, complete seal, and cool.

With steam-pressure outfit sterilize 60 minutes at 5 to 10 lb. pressure.

Pumpkin—Winter Squash.

Remove seeds. Cut the pumpkin or squash into strips. Peel and remove stringy centre. Slice into small pieces and boil until thick. Pack in jar and sterilize 120 minutes in hot-water bath.

With steam-pressure outfit sterilize 60 minutes at 5 to 10 lb. pressure.

String Beans.

Wash and remove ends of strings. Blanch from 5 to 10 minutes, depending on age. Cold-dip. Pack immediately in jar, placing pods lengthwise in jar. Add 1 teaspoon salt and cover with boiling water. Put on rubber and adjust top. Sterilize 120 minutes in hot-water bath. Remove, complete seal, and cool.

With steam-pressure outfit sterilize 60 minutes at 5 to 10 lb. pressure.

Tomatoes.

Select tomatoes that are ripe, but not overripe, free from blemishes, and of medium size if possible. They should be red to the stem end, since green parts produce poor flavour and colour. Imperfect tomatoes may be used for catsup and purée, or made into a hot liquid and used for filling in spaces left in a jar after it is packed with whole tomatoes. Scald a few tomatoes at a time in boiling water for from ½ to 2 minutes, using a wire basket or a thin cloth. Dip them into cold water and remove them quickly. With a small sharp paring-knife cut out the stem-core; then with a quick turn of the wrist twist the skins from the tomatoes without removing the pulp. If the pulp adheres to the skin, the tomatoes have been scalded too long or not long enough. In packing the scalded tomatoes into the jars, pressing them down firmly with a wooden spoon, fill the jars to within ¼ inch of the top with boiling tomato-juice. Add 1 teaspoon salt for each quart and from 1 teaspoon to 1 tablespoon sugar if desired. Put on rubber and adjust top. Sterilize 22 minutes in hot-water bath. Remove, complete seal, and cool.

With steam-pressure outfit sterilize 15 minutes at 5 to 10 lb. pressure.

Swiss Chard.

Cut white stalks from leaves. Can each separately and according to the method described for greens. The stalks, however, should be blanched in boiling water instead of in steam.

DIRECTIONS FOR CANNING MEATS BY THE COLD-PACK METHOD.

Meats may be canned as successfully as fruits and vegetables if proper methods are employed. The Fractional or Intermittent Sterilizing Method should not be used, as this particular method allows the development and reproduction of the bacteria producing ptomaine poisoning.

METHOD 1.

Free the meat from the bone and cut it in pieces of such sizes that they will go into the jars easily. If additional flavouring is desired, sear and brown the meat quickly in hot fat in a frying-pan, but do not cook it through. Pack the raw meat solidly in tested clean glass jars, filling the jar to within ¾ inch from the top. Sprinkle the top of the meat with ½ teaspoon salt for each pint of meat. Add no water. Celery-leaves, onion-juice, or other seasonings may be added if desired. Adjust on the jar a new rubber of good quality. Place the cover on the top of the jar and adjust but do not fasten the upper wire clamp, or, if a Mason jar is used, partly screw on the cover. Place the jars in a sterilizer in which there is warm water which covers the tops of the jars about 1 inch. Sterilize the meat by cooking it from 4 to 5 hours, beginning to count the time when the water around the jars reaches the boiling-point. Keep the water jumping. Before removing the cans from the sterilizer complete the sealing of each jar by adjusting the lower wire of the clamp, or, in the case of a Mason jar, by screwing the top tight. Do not invert the jar while it is cooling. If the jar were inverted, the fat, which is lighter, would rise to the bottom of the jar and cool and harden there. When the jar is left upright the fat comes to the top of the jar and hardens there, forming an extra seal.

METHOD 2.

Sear the meat in a hot oven in hot fat or in boiling water, and steam it or simmer it until it can be torn apart. Pack the meat into the jar; fill the space with stock which is made by boiling the broken bones and skin in water until the stock will form a jelly when cooled. Add ½ teaspoon salt to each pint of meat. Sterilize the meat for

3 hours as in Method 1. Unless the meat is first browned it does not have so good a flavour as that of raw meat steamed in the can.

Canned Chicken.

Chicken may be successfully canned by either of the two methods suggested. A fowl weighing 2 lb. when dressed should make a pint can of solid meat and a pint of stock thick enough to jelly. A fowl weighing 3 lb. should fill 1½ pint cans.

Chicken Stock.

All bones and trimmings of the chicken should be covered with cold water, salted, and allowed to stand overnight. Slowly simmer until flesh drops in shreds from the bones and the liquid or stock is concentrated. Seasonings such as grated onion and a bit of celery-leaf may be added. Strain the stock if desired, reheat it, and boil it for 10 minutes. Pour it into sterilized jars, and sterilize it as described in Method 1.

CURING OF FISH.

A manual on the curing of fish has been issued by the Provincial Fisheries Department (Bulletin No. 2), and copies may be had upon application to the Provincial Fisheries Department, Victoria, B.C.

DIRECTIONS FOR CANNING FRUIT BY THE COLD-PACK METHOD.

(1.) Select well-grown, firm, and not overripe fruit.

(2.) If possible, can fruit on the day it is picked.

(3.) Wash, pare, or otherwise prepare the fruit; remove all bruises or decayed parts.

(4.) If there is much variation in size, grade the fruit so that the contents of each jar will be as nearly uniform as possible.

(5.) Blanch or scald in boiling water a small quantity of the fruit at a time. The number of minutes required for blanching is given in table. Do not blanch cherries, berries, or plums.

(6.) Chill the outside of the blanched fruit by immersing it for a brief period in a large vessel of cold water. Do not attempt to cool the fruit thoroughly by this cold-dip.

(7.) Pack the fruit firmly in clean, tested jars to within ½ inch of the top.

(8.) Fill the jars to within ¼ inch of the top with boiling syrup or hot water.

(9.) Place a new rubber on each jar, adjust the cover of the jar, and partially seal it.

(10.) Sterilize the jars for the required length of time. The jars should be immersed in sufficient boiling water to cover the tops to the depth of about 1 inch. Do not begin to time the sterilization until the water boils rapidly. Keep the water boiling during the sterilization period. Remove the jars from the sterilizer, seal them, and invert them to cool. Avoid a draught on the jars, but cool them as rapidly as possible.

(11.) Wash the jars thoroughly, label them, and set them away. Store red fruits in a dark place to prevent loss of colour.

CANNING FRUIT WITHOUT SUGAR.

Fruit may be canned in water instead of syrup without in any way affecting the ease of canning, the keeping quality of the fruit, or the wholesomeness of the product. When a sugar shortage existed this method enabled housewives to conserve the fruit surplus. In times when sugar is plentiful this method is not advisable, as sugar adds to the attractiveness of texture and flavour of fruits preserved with it.

CALIFORNIA SYRUP FORMULA.

The following syrups are made according to the California Syrup Formula. The amount used depends upon the ripeness of the fruit and the fullness of the pack.

(1.) *Very Thin Syrup.*—One cup or less sugar to 1 cup water and heat to the boiling-point.

(2.) *Thin Syrup.*—One and one-half cups sugar to 1 cup water and heat to the boiling-point.

(3.) *Medium Thin Syrup.*—Make in the proportion of 1½ cups sugar and 1 cup water; boil 2 or 3 minutes or until the solution begins to be syrupy.

(4.) *Medium Thick Syrup.*—Make in proportion of 1½ cups sugar and 1 cup water; boil 5 minutes.

(5.) *Thick Syrup.*—One and one-half cups sugar and 1 cup water; boil 8 to 12 minutes or until it forms a soft ball in water.

NOTE.—In making syrups see that the sugar is thoroughly dissolved before the syrup is allowed to boil, else the sugar is apt to candy in the bottom of the jar.

Apples.

Pare, core, then blanch and cold-dip. Pack in jars and add syrup. Boiled cider may be used instead of syrup if desired. Sterilize the required length of time.

Cherries.

Stone the cherries; and, if sour cherries, allow ½ lb. sugar to every pound of cherries. If sweet cherries, ¼ lb. sugar. Put the cherries and sugar in layers in a porcelain-lined kettle, let stand 1 or 2 hours, then place over a moderate fire and bring to boiling-point. Skim and can immediately in hot sterilized jars. Seal and set away to store.

Pears.

The Bartlett pear is the best variety for canning. Remove skins, cut in halves, or other desired shape. They may be canned whole with the stems on. Pears discolour rapidly as soon as peeled, and for this reason they should be dropped into cold water and allowed to stand until ready to can. Pack in jars, add syrup, and sterilize the required length of time.

Apples with Pinapple.

4 lb. apples.   1¼ lb. sugar.
1 good-sized pineapple.   1 quart water.

Pare, core, and quarter the apples. Pare the pineapple, and with a silver knife carefully remove the eyes, then grate it. Cover the apples with boiling water, bring quickly to a boil, then simmer gently for 5 minutes. Put the sugar and water in another kettle, stir constantly until the sugar is dissolved, then add the grated pineapple, and bring the whole to boiling-point. Lift the apples, drain and slide them carefully into the syrup. Simmer until the apples are tender, and can in hot sterilized jars. Seal and set away to store.

Peaches.

Blanch, cold-dip, then peel, cut in halves, and remove stones. Procure meat from pits. Add to fruit and pack in jars. Cover with syrup. Adjust covers and sterilize the required length of time.

"RAW CANNING" OF FRUITS.

Method 1.

Small fruits like raspberries, strawberries, or sliced peaches can be sterilized so as to retain their shape and colour and natural flavour without actual cooking. Pack fruit into sterilized jars. Make a medium thick syrup, and while it is boiling pour it over the fruit and seal tightly. Put the jars in a kettle or wash-tub and fill the vessel covering tops of the jars with boiling water, cover closely, and allow the jars to remain until the water is cold.

Method 2.

It has been found that strawberries and raspberries may be canned without subjecting the jars to further heat after they have been filled with the boiling syrup. If this method is used, select freshly picked fruit which is sound and firm. Strawberries should be first washed, then hulled, so as to prevent loss of juice and colour. Fill sterilized hot glass jars with the berries and make a firm pack. Fill jar to overflowing with a medium thick syrup which is boiling hot (the directions for which are given). With a sterilized fork lift up the fruit, allowing the air to escape, and refill with syrup. Seal and clamp immediately.

NOTE.—A lighter syrup should not be used, as there is not always sufficient heat in it to sterilize the fruit. Raspberries and strawberries usually spoil because of the presence of yeasts and moulds. These spores are on the surface of the fruits and so are destroyed by the boiling hot syrup.

Berry Preserves.

Wash, drain, and hull the berries. Add an equal weight of sugar. Crush the berries with the sugar and mix them well. Allow the berries to stand for 24 hours, stirring them occasionally until the sugar is dissolved. Seal them in glass jars and keep the jars in a cool, dark place. Strawberries and raspberries canned in this way are excellent for shortcake. Red currants may be canned in the same manner.

Sun Preserves.

Method 1.

Fruits that lend themselves especially well to the following method of preserving are strawberries, cherries, white currants, and raspberries. Use a pound of sugar to each pound of fruit. Put a layer of fruit in the bottom of a preserving-kettle and add 1 or 2 tablespoons water. Alternate the layers of sugar and fruit. Heat the mixture carefully until the sugar is dissolved; avoid crushing the fruit if possible. Boil the moisture for from 5 to 7 minutes. Pour it on to large platters and set it in the sun for a day. It should thicken or jelly on the platter. After it is cold and thickened, transfer from the platter to sterilized jars and seal or cover at once with paraffin.

Method 2.

Fruits that lend themselves especially well to the following method of preserving are peaches, apricots, raspberries, and plums. Carefully wipe or pick over the fruit to be preserved. Cut peaches, plums, or apricots in halves and remove the pits. Spread the fruit on racks or boards and set it in the sun to dry for one or two days. The fruit should not be left out overnight to gather moisture. Weigh the fruit and use a pound of either brown or white sugar to each pound of fruit. Pack alternate layers of fruit and sugar in jars, being careful to have the top layer of sugar. The sugar will dissolve gradually and form a thick rich syrup around the fruit. The mixture should be kept covered, but need not be sealed.

PRESERVES, CONSERVES, ETC.

Grape Preserves.

Weigh the grapes and allow ¾ lb. sugar to 1 lb. fruit. Rinse the bunches of grapes in cold water, drain, and squeeze the pulp from the skin of each grape. Heat the pulp gradually and cook until the seeds come out easily; 10 or 15 minutes will be required. Pass through a sieve just fine enough to keep back the seeds. Cook the skins and the pulp 10 minutes, then add the sugar and continue cooking until the liquid thickens slightly. Store in earthen or glass jars.

Grape Conserve.

3 lb. seeded grapes.   3 lb. sugar.
1 lb. English walnuts (broken into small pieces).

Measure the ingredients and cook them together as for jam. The juice of one orange and the peel of half an orange cut in small pieces may be added for variation.

Tomato Preserves.

1 peck tomatoes (chopped).   6 lb. sugar.
4 or 6 lemons (sliced thin).

Cook the mixture until it is thick and clear, pour it into sterilized jars, and seal.

Orange Marmalade.

12 thin-skinned oranges.   3 lemons.

Wash and slice the fruit as thin as paper or grind it fine. To every quart of fruit add 1½ quarts of water and let the mixture stand overnight. In the morning cook it slowly until tender, about 3 hours, leaving the lid off. Measure the cooked fruit and add an equal amount of hot sugar. Cook the mixture until a drop or two will jell on a cold plate, about 10 minutes. If bitter marmalade is desired, use about 6 bitter oranges and 6 sweet oranges.

Amber Marmalade.

Take one, each, grapefruit, orange, and lemon; wash and wipe and cut in quarters; then thorough peel and pulp, cut into thin slices, discarding seeds. Add 7 pints cold water and let stand overnight. Cook with the cover off until the peel is tender. It will take several hours. Set aside overnight. Heat again to the boiling-point and add 10 cups (5 lb.) of hot sugar, and cook, stirring occasionally until the syrup thickens slightly on a cold dish, or when the marmalade is put in a glass the peel will remain distributed from bottom to top of glass.

Citron, Melon, or Watermelon Preserves.

Remove the green outer rind of the melon and cut the remainder in pieces of small size. Cover with cold water and add a tablespoon of salt for each quart of water. Let stand overnight in the salted water; then drain and rinse thoroughly. Cook in boiling water until transparent. Drain carefully. For each pound of rind make a syrup of ¾ lb. sugar and ½ cup water and skim thoroughly; then add the melon, and for each pound add ½ oz. ginger-root and 2 tablespoons vinegar or 1 lemon cut in slices; a few sticks of cinnamon may be added to the syrup. Cook about 20 minutes, or until the pieces of melon look rich and full. Skim from the syrup into jars; boil the syrup until rich and thick and pour over the fruit in the jars.

Pear Chips.

8 lb. pears.	4 lb. sugar.
¼ lb. preserved ginger.	4 lemons.
½ tablespoon ground ginger.	

Select pears which are firm and not overripe. Remove stems, wipe, quarter, and core; then chip into small pieces (but do not remove skins). Add sugar and ginger and let stand overnight. In the morning add lemons cut in small pieces, rejecting seeds, and cook slowly 3 hours or until thick. Put into marmalade-glasses and cover with paraffin.

Quince Honey.

Pare and grate 5 large quinces. To 1 pint boiling water add 5 lb. sugar. Stir over fire until sugar is dissolved, but do not allow the syrup to boil until the sugar is thoroughly dissolved. When boiling, add quince and cook 15 or 20 minutes. Turn into glasses. When cold it should be about the colour and consistency of honey.

Spiced Gooseberries.

4 pints partially ripe gooseberries.

3 lb. brown sugar.	1 teaspoon salt.
1 cup vinegar.	⅛ teaspoon cayenne.
1 teaspoon whole cloves.	1 teaspoon whole cloves.
Few sticks of cinnamon.	½ oz. ginger-root.

Tie spices in a bag. Cook vinegar and sugar 5 minutes. Add spice and remaining ingredients and cook slowly 1 hour. Keep in stone jar or glasses.

Spiced Currants.

4 lb. currants.	2 tablespoons cloves.
2 lb. brown sugar.	1 teaspoon salt.
2 tablespoons cinnamon.	1 cup vinegar.

Remove stems and wash currants. Add the remaining ingredients and boil 20 minutes. Keep in stone jar or glasses.

Apple Butter.

1 bushel apples.	8 quarts cider.

Cover and boil until tender. Rub the pulp through a strainer and cook 30 minutes longer; then measure. For each gallon add 8 cups sugar, 2½ tablespoons ground cloves, and 2½ tablespoons ground cinnamon. Stir and boil 20 minutes longer; fill into jars and seal with paraffin.

Apple Butter without Sugar.

1 bushel sweet apples.	8 quarts cider.

Cook until tender; put through a strainer and cook until thick. Add 3 tablespoons ground cloves and 3 tablespoons cinnamon. Give 3 to 4 hours' slow boiling; fill into jars and seal with paraffin.

Red Currant Conserve.

5 lb. currants.	2 lb. raisins
5 lb. sugar.	(chopped finely).
5 large oranges (use the rind of 3 boiled and chopped and the juice of all).	

Boil mixture for 25 minutes or a little longer. Seal in glasses.

Raspberry and Currant Preserve.

6 quarts currants.	6 lb. sugar.
8 quarts raspberries.	

Pick over, wash, and drain currants. Put into a preserving-kettle, adding a few at a time, and mash. Cook 1 hour; strain through double thicknesses of cheese-cloth. Return to kettle, add hot sugar, heat to boiling-point, and cook slowly 20 minutes. Add 1 quart raspberries when syrup again reaches the boiling-point. Skim out raspberries, put in jar, and repeat until raspberries are used. Fill jar to overflowing with syrup and screw on tops.

Mint Jelly.

The best mint jelly is made with the juice of underripe apples as a basis. Wash fresh mint-leaves thoroughly. To 1 cup mint-leaves (packed solid) add 1 cup boiling water; set the mixture on the back of the stove and steep it for 1 hour. Lay a piece of cheese-cloth over a bowl, pour the steeped mint-leaves into it, twist the ends of the cheese-cloth, and press out all moisture. To 1 cup concentrated apple-juice add 2 tablespoons mint-juice. If the mint flavour is not sufficiently pronounced, add a drop or two of mint extract. Use ¾ cup hot sugar to each cup of juice, and boil the mixture rapidly until the jelly test can be obtained. Just before it is poured into the scalded glasses colour it green with vegetable colouring-matter.

Cider Apple Sauce.

Reduce 4 quarts cider to 2 by boiling. Add enough pared, cored, and quartered apples to fill a good-sized kettle. Cook slowly for 4 hours. Pour into jars and seal or keep in stone jars.

Raspberry or Strawberry Jam.

3 lb. raspberries or strawberries.
6 cups sugar.   3 cups water.

Put sugar and water in porcelain kettle on stove. Stir and dissolve sugar, not allowing it to boil until sugar is dissolved. Boil until syrup will hair well from spoon. Add fruit and boil quickly from 15 to 30 minutes, stirring often, or until fruit-juice will jelly. Remove from heat, partially cool, and skim. Pour in stone jars or jelly-glasses.

NOTE.—If the berries are put into the hot, thick syrup and cooked quickly, more of their natural colour is retained. Better results are obtained when made in small amounts.

Black, Red, or White Currant Jam.

4 lb. currants.   4 lb. sugar.
½ cup water or apple-juice.

Stem currants, add to apple-juice, and boil, stirring often until currants are cooked. Heat sugar and add to fruit. Stir until dissolved and boil 5 minutes or until thick. Turn into glasses and cover.

Damson Jam.

4 lb. damsons.   4 lb. sugar.

Damsons make the best jam and jelly if they are the real English damsons. Very few are grown in Canada. They are a small sour blue plum and when cooked are of a rich red colour. Wash fruit; place in kettle with a little water. Cook until plums are soft enough to remove pits. Pits may be removed if desired. Heat sugar and add to fruit; stir until dissolved and boil 3 minutes. Turn into glasses and cover. The jam should be jellied and be of a rich red colour.

JELLY-MAKING.

In order to make good jelly, fruit-juice must contain two ingredients, acid and pectin. The pectin is generally known as a substance in fruits which makes jelly "jell." It is found in the largest quantities in the cores, seeds, and hard parts of fruit, and as the fruit ripens it is changed into a substance which has very little of the jellying property. For this reason it is important to use fairly ripe fruit and to include cores, seeds, and skins in the first boiling to extract the juice.

Test for Pectin.—To test fruit-juice for pectin, add 1 teaspoon ordinary alcohol to 1 teaspoon cold fruit-juice. If pectin is present, a solid mass, which is pectin, collects. This indicates that in making jelly 1 part of sugar should be used to 1 part of juice. If there is no pectin, the solution should remain clear.

The changing of the juice from a liquid to a jelly is brought about by the combined effect of sugar, acid, and boiling upon the pectin of the fruit-juice. Some fruits, such as peaches, quinces, pears, and sweet apples, contain sufficient pectin but are deficient in acid, and when making jelly from these fruits lemon-juice is added. A fruit that jells with difficulty may be combined with one that jells readily; apples, though possessing little flavour, have all the necessary jellying qualities. When any desired flavour is added, good jelly results. Fruits suitable for jelly-making are currants, ripe and partially ripe grapes, crab-apples, sour apples, green gooseberries, wild cherries, and plums. Raspberries may be used, though they jell less rapidly.

It is, of course, possible to supply the deficiency of either acid or pectin. In oranges and lemons the white material between the pulp and yellow rind is very rich in pectin. This may be extracted by grinding or chopping fine the thick white part, soaking in cold water 12 to 24 hours, and then simmering 1 hour. Equally good results may be obtained, however, by adding a generous supply of apple cores and skins to the fruit before boiling to extract the juice. A deficiency of acid may be likely overcome by adding some acid fruit. Rhubarb-juice added to any fruit-juice will bring out the flavour and add zest to the jelly. Tartaric or citric acid are perfectly safe fruit products and may be obtained in crystalline form. One level teaspoon to a quart of juice is usually sufficient; however, this depends on the acidity of the fruit. To test, stir the juice until all acid crystals are dissolved; then taste. It should be about as acid as good tart apples.

Apple Jelly.

Procure apples that are a little underripe. Wash and cut into pieces without peeling or removing the cores and seeds. Put them into a kettle, just cover them with cold water, and cook them until they are soft and tender. Transfer them to a jelly-bag and let them drain. Carefully avoid applying pressure if clear jelly is required. When the juice has all drained out, measure it and return to the kettle. Allow it to boil for 10 minutes. For each measure of juice add 1 measure of hot sugar. Add sugar gradually, and when it is thoroughly dissolved allow jelly to boil about 5 minutes, when it should jell on a cold plate. Pour into jelly-glasses and cover it with melted paraffin.

NOTE.—Sweet geranium, peach, or mint leaves are often placed on the surface of hot jelly before the paraffin is used. They give a delicious flavour which is liked by many.

Quince and Apple Jelly.

Remove the fuzz from the quinces with a damp cloth. Cut into small pieces and for every 2 measures of quinces add 1 measure of apples. Put them into a preserving-kettle. Cover them with water and boil them until they are soft. Proceed according to the direction given for apple jelly.

Currant Jelly.

Do not gather the currants just after a rain. Extract the juice by pressing a few at a time in a cloth; then let drain through a cloth without pressure. Let as many cups of sugar as there are of juice heat in the oven without discolouring (stir often); then when the juice boils add the sugar and let boil a few minutes or until a little will jell on a cold plate. Cook but a small portion of juice at a time.

Currant Jelly, also Grape, Blackberry, Plum, Apple, etc.

Cut apples in quarters without removing skins or cores unless defective, cut plums in halves, and pull grapes and currants from the stems. A little water needs be added to apples and crab-apples, quinces, and other dry fruit. Avoid the use of water with currants, grapes, etc.; let cook until the pulp is soft; then drain without pressure. Press the bag to get the last of the juice, and with this make a second quality of jelly. Jelly made of dry fruits will harden if it is set aside in the glasses, and if cooked to the consistency usually desired in currant and similar jelly will be too firm and solid in a few weeks. Equal quantities by weight of blackberries and apples make delicious jelly. Green gooseberries or green plums give a delicately tinted jelly.

Cranberry Jelly.

Pick over and wash 4 cups cranberries. Put in a stew-pan with 2 cups boiling water and boil 20 minutes. Rub through a sieve, add 2 cups sugar, and cook 5 minutes. Turn into a mould or glasses. If boiling water is used more of the original colour is retained.

PICKLING.

Pickling is an important branch of preparedness for the winter months. Vegetables and fruit are pickled and preserved by the use of wholesome preservatives, such as salt, vinegar, spices, and sugar. If in the making of pickles these preservatives are used in reasonable amounts they will be a wholesome addition to a meal. Pickles give flavour to a meal and so stimulate the flow of digestive juices; but there are pickles and pickles, and those made strong with vinegar and very highly seasoned with spices should be eaten sparingly and never given to children.

In pickling, vegetables are usually soaked overnight in a brine made of 1 cup salt and 1 quart water. This brine removes the water of the vegetable and so prevents weakening of the vinegar. In the morning the brine is drained off.

Alum should not be used to make the vegetables crisp, as it is harmful to the human body. A firm product is obtained if the vegetables are not cooked too long or at too high a temperature.

Enamelled, agate, or porcelain-lined kettles should be used when cooking mixtures containing vinegar. Pickles put in crocks should be well covered with vinegar to prevent moulding. Cucumber pickles are sometimes coloured green with sulphate of copper, which is a deadly poison. They are tinted green also by scalding in a brass or copper kettle; but while pickles so coloured might not cause illness, no one who pays any regard to health would venture to eat them.

Grape-leaves and cabbage-leaves are said to help in retaining the natural green colour of cucumbers and unripe tomatoes. The bottom and sides of the kettle are lined with the leaves, the kettle is then filled with the mixture to be pickled, and the top of the mixture is covered with leaves.

Cold Tomato Relish.

Eight quarts firm ripe tomatoes; scald, cold-dip, and then chop in small pieces. To the chopped tomatoes add:—

2 cups chopped onion.	4 chopped peppers.
2 ,, chopped celery.	1 teaspoon ground mace.
2 ,, sugar.	1 ,, black pepper.
1 cup white mustard-seed.	4 teaspoons cinnamon.
½ ,, salt.	3 pints vinegar.

Mix all together and pack in sterilized jars.

Cucumber Pickles.

Soak in brine made of 1 cup salt to 2 cups water for a day and a night. Remove from brine, rinse in cold water, and drain. Cover with vinegar, add 1 tablespoon brown sugar, some stick cinnamon, and cloves to every quart of vinegar used; bring to a boil and pack in jars. For sweet pickles use 1 cup sugar to 1 quart vinegar.

Chili Sauce (1).

2 doz. ripe tomatoes.
6 peppers (3 to be hot).
3 onions.
¼ cup sugar.
2 tablespoons salt.
1 teaspoon each of cloves, nutmeg, and allspice.
1 quart vinegar.

Simmer 1 hour. Pour into sterilized jars or bottles and seal while hot.

Catsup.

Two quarts ripe tomatoes. Boil and strain. Add 2 tablespoons salt, 2 cups vinegar, 1 cup sugar, and 1 level teaspoon cayenne pepper. Boil until thick. Pour into hot sterilized bottles. Put the corks in tight and apply hot paraffin to the tops with a brush to make an air-tight seal.

Beet Pickles.

Twist leaves and stalks from beets. Wash but do not cut roots nor tops. Cover with boiling salted water and boil until tender, but not soft. Dip in cold water. Rub off skins and trim. If beets are small, leave whole. Cut larger ones in halves or quarters, cutting lengthwise so as to retain as much of the juice and colour as possible. Fill hot sterilized jars with the beets and cover with a boiling hot pickle made of the following ingredients:—

1½ cups vinegar.
1½ " water.
1 cup brown sugar.
¼ cup salt.
¼ teaspoon cayenne pepper.

Fill jars to overflowing. Seal and set away.

Celery Relish.

1 doz. heads celery.
3 onions.
½ lb. mustard.
¼ " mustard-seed.
4 cups white sugar.
2 quarts vinegar.
1 tablespoon tumeric (level).
2 tablespoons salt.
½ teaspoon cayenne pepper.
3 tablespoons flour.

Mix ingredients, omitting the flour and tumeric, and simmer for 2 hours. Mix flour and tumeric together with a little vinegar or water. Add to pickle and cook 5 minutes longer. Bottle and seal.

Rhubarb and Onion Pickle.

2 quarts rhubarb (cut in small pieces).
2 quarts minced onion.   1½ pints vinegar.
Cook these together 20 minutes and add:—

4 lb. light-brown sugar.   1 tablespoon allspice.
1 teaspoon pepper.   ½   „   cloves.
1 tablespoon salt.   A piece of ginger-root.
1   „   cinnamon.   A little mustard-seed.

Boil until fruit is soft. Bottle and seal. (NOTE.—This makes about 7 pints.)

Cherry Olives.

Cherry olives is a relish which is characteristic of the Kootenay District of British Columbia. The cherries are prepared and served as olives. The Royal Anne is considered one of the most suitable varieties.

Prepare cherries by washing and clipping off a portion of the stem. Pack them in jars and cover them with the following solution:—

1 pint vinegar.   2 tablespoons salt.
1 pint water.

NOTE.—No heating, no cooking; keep in a cool place.

Cherry Relish.

Remove the pits from cherries and drain them. Cover them with a vinegar solution made in the proportion of ¾ cup vinegar to 1 quart water. After 5 or 6 hours drain the cherries, weigh them, and add an equal weight of sugar. Allow the cherries to stand overnight. Seal them in glass jars and keep them in a cool dark place. The vinegar solution that has been drained off may be used in making various kinds of sweet pickles.

Mustard Pickles.

(MacDonald Institute, Guelph.)

2 quarts cucumber.   4 cups brown sugar.
2   „   small silver onions. ¼ lb. mustard-seed tied in
1 head cauliflower.   muslin bag.
3 green peppers.   ¼ „ ground mustard.
3 red peppers.   ½ oz. tumeric.
½ gallon best cider vinegar. 1 cup flour.

Wash and prepare vegetables; put each kind to soak separately in strong hot brine which will float an egg (1 cup salt to 2 cups water).

Let stand overnight. Drain and rinse in cold water. Put the vinegar, sugar, and spices together and heat to boiling-point. Add vegetables and scald for 5 minutes. Lift vegetables into sterilized jars; thicken the vinegar with flour and mustard which has been mixed to a paste with cold vinegar. Stir vinegar while thickening and cook 5 minutes; then pour over vegetables into jars. Seal.

Ripe Cucumber Pickle.

(MacDonald Institute, Guelph.)

12 large ripe cucumbers.
6 to 12 onions.
3 pints vinegar.
2 cups sugar.
4 tablespoons flour.
¼ cup mixed whole spice, cloves, and allspice.
1 tablespoon ground mustard.
1 „ curry-powder.

Wash and peel cucumbers, taking out seeds and pith. Add onions and sprinkle well with salt and let stand overnight. Drain and rinse in cold water. Boil vinegar, sugar, and spice together in a bag. Add vegetables and let scald 10 minutes. Remove vegetables to bottles and thicken vinegar by adding the flour, mustard, and tumeric, which have been mixed to a paste with a little cold vinegar. Pour over the vegetables and seal.

Sweet Pickle (Gherkin).

1 quart vinegar.
2 tablespoons mustard-seed.
2 „ whole allspice.
2 „ peppercorns.
2 pieces ginger-root.
2 blades mace.
Some small pieces of horse-radish.
2 cups brown sugar.
1 teaspoon cloves.
A stick of cinnamon may be added if desired.

Prepare the vegetables, soak in brine overnight, drain and rinse. Heat vinegar, sugar, and spices together. Let boil for 5 or 10 minutes; put vegetables in hot sterilized jars and fill to overflowing with the vinegar. Seal.

Chili Sauce (2).

35 large tomatoes.
4 small red peppers.
10 average-sized onions.
6 cups vinegar.
2½ cups sugar.
5 tablespoons salt.

Chop onions and peppers together, add to other ingredients, and boil 3 hours. Bottle and seal.

PICKLING.

Pickled Beans.

Five lb. butter beans, cut small and cooked for ½ hour in salt water. Mix together 1 cup flour, 1 cup mustard, 3 lb. brown sugar, 2 tablespoons celery-seed, and 2 tablespoons tumeric. Add 3 pints vinegar, stir until smooth, and boil 5 minutes. Add beans and bring to the boiling-point. Bottle and seal.

Chow Chow or Piccalilli.

1 medium cabbage.
1 cauliflower.
3 quarts onions.
½ doz. cucumbers (ripe or green).
2 heads celery.

NOTE.—Red or green tomatoes may be added if desired. One or two green or red peppers. Prepare vegetables and chop each fine. Sprinkle with salt and let stand overnight. Drain off liquid. Make a sauce as follows:—

3½ quarts vinegar.
3 or 6 cups sugar.
6 tablespoons mustard.
2   ,,   tumeric.
¾ cup flour.

Add cooked sauce to the vegetables and cook 10 to 15 minutes. Bottle and seal.

Corn Relish (1).

1 small cabbage.
1 large onion.
6 ears corn.
2 tablespoons salt.
2 tablespoons flour.
1½ cups brown sugar.
2 hot peppers.
1 pint vinegar.
1½ tablespoons mustard.

Steam corn 30 minutes. Cut from the cob and add to the chopped cabbage, onion, and peppers. Mix the flour, sugar, mustard, and salt; add the vinegar. Add mixture to the vegetables and simmer 30 minutes. Pour into sterilized jars or bottles and seal while hot.

Corn Relish (2).

Cut the corn from 2 dozen ears; chop rather fine 1 head cabbage, 4 large onions, 4 green peppers, and 1 red pepper, first discarding the seeds of the pepper; add 1 quart vinegar and set to boil. Mix together 3 cups sugar, ¾ cup flour, ½ cup salt, ¼ cup dry mustard, and 1 teaspoon tumeric; when well mixed stir in 1 quart vinegar, and then stir the mixture into the hot vegetables. Let simmer ½ hour; add 2 teaspoons celery-seed and store as canned fruit or vegetables.

Table Relish.

Chop:—
    4 quarts cabbage.   6 large onions.
    2 „ tomatoes (1 quart to be green).   2 hot peppers.

Add:—
    2 oz. white mustard-seed.   ¼ cup salt.
    1 oz. celery-seed.   2 lb. sugar.
              2 quarts vinegar.

Simmer 1 hour. Pour into sterilized jars or bottles and seal while hot.

Apple Chutney.

    12 sour apples.   ½ cup currant jelly.
    1 onion.   2 cups sugar.
    3 peppers (1 red).   The juice of 4 lemons.
    1 cup seeded and chopped raisins.   1 tablespoon ground ginger.
          ¼ teaspoon cayenne.
    1 pint cider vinegar.   1 tablespoon salt.

Chop the apples, onions, and peppers very fine, add the vinegar and jelly, and let simmer 1 hour, stirring often. Add the other ingredients and cook another hour, stirring constantly. Store as canned fruit.

Cranberry Sauce.

Pick over and wash 3 cups cranberries. Put in stew-pan with 1 cup boiling water. Boil 10 minutes; then add 1½ cups sugar. Stir until dissolved.

NOTE.—When cranberries are put into boiling water they keep their colour better.

French Pickle.

Chop fine ½ peck green tomatoes, 1 head cauliflower, 15 white onions, and 10 large green cucumbers. Put a layer of vegetables into a porcelain dish and sprinkle with salt; continue the layers of vegetables and salt until all are used; let stand overnight; then drain, discarding the liquid. Heat 3 quarts cider vinegar, 3 lb. brown sugar, ¼ cup tumeric, ¼ cup black pepper-seed, 1 oz. celery-seed, ¾ cup mustard-seed, and 3 red peppers chopped fine. Heat to the boiling-point and pour over the vegetables. Let stand overnight; then drain the liquid from the vegetables, reheat, and again pour over the vegetables. Repeat this process the third morning; then when the mixture becomes cold, stir into it ¼ lb. ground mustard and 1 teaspoon curry-powder mixed with 1 cup olive-oil and 3 cups vinegar (use less mustard and vinegar if desired).

Sweet Pickled Peaches.

½ peck peaches.
2 lb. brown sugar.
1 pint vinegar.
½ cup water.
1 oz. stick cinnamon.
½ oz. whole cloves.

Boil sugar, vinegar, and cinnamon 20 minutes. Dip peaches quickly in boiling water; then rub off the skins. Stick each peach with 2 or 3 cloves. Put into syrup and cook until soft, using one-half peaches at a time. Pack in sterilized jars and add syrup. Seal.

Sweet Pickled Pears.

Follow recipe for sweet pickled peaches, using pears in place of peaches.

Beet Relish.

1 cup chopped cold cooked beets.
3 tablespoons grated horse-radish root.
2 tablespoons lemon-juice.
2 teaspoons powdered sugar.
1 teaspoon salt.

Mix ingredients in order given. Canned beets may be used instead of fresh ones, and bottled horse-radish if of strong flavour and well drained. This is delicious served with cold meat or fish.

Dill Pickles.

Cover cucumbers of medium size with clear water. Next day drain and wipe dry. Pack in fruit-jars, using plenty of fresh dill between. To each ½-gallon jar add 2 small red peppers, 2 bay-leaves, and 2 thin slices horse-radish root. To 6 quarts water add 1 lb. rock salt. Heat mixture to boiling-point; add 1 quart vinegar. Pour at once over cucumbers, covering them well. Seal tight while hot. Dill pickles will keep in covered kegs or crocks without being sealed. Dill is a hardy plant of medicinal value and may be successfully grown in any vegetable-garden.

A Time-table for Canning Vegetables which are most safely canned by Three Periods of Sterilization or the So-called Intermittent Method.

(From Cornell Reading Course for the Farm Home.)

Food.	Time of Blanching.	Time of Cooking.	
		First Day.	Second and Third Days.
	Minutes.	Minutes.	Minutes.
Asparagus	5	60	60
Beans	5	60	60
*Beets	6-10	60	60
*Carrots	6-10	60	60
*Cauliflower	5	60	60
Corn	5-10	60-75	60-75 (*depending on closeness of pack*).
*Parsnips	6-10	60	60
Peas	5	60	60
*Pumpkin		75	75
Spinach and other greens	5	75-90	75-90 (*depending on closeness of pack*).
*Squash		75	75
*Succotash		60	60

* Those vegetables marked with a star are not as difficult to sterilize as the others, and the risk of canning them by the continuous method is therefore less than with the unstarred list.

A Time-table for Canning Fruits, Acid Vegetables, and Meats by the Single or Continuous Period of Sterilization.

Food.	Time of Blanching.	Time of Cooking.	
		If the Hot-water Bath is used.	If the Pressure Cooker is used, 5 lb.
	Minutes.	Minutes.	Minutes.
Apple cider	1-2	20	12
Apples	1-2	20-30	10
Apricots	1-2	16	10
Blackberries, dewberries		16	6
Cherries		16	10
Fruit juices		20	10
Grapes, plums		16	10
Huckleberries		16	8
Peaches	1-2	16	10
Pears	1-2	20	10
Pineapples		60	40
Quinces	1-2	60	40
Raspberries		16	8
Rhubarb	1-2	16	10
Strawberries		16	10
Sauerkraut		60	50
Tomatoes	1-2	22	10
Tomatoes and corn		90	60
Tomato-juice		20	15
Meat		300	180

A Time-table for Canning Vegetables where the Pressure Cooker is used, or where the Owner is willing to take the Risk of the Uncertain Single Period of Sterilization in the Water-bath.

Food.	Time of Blanching.	Time of Cooking.		
		When the Hot-water Bath is used (a Risky Method).	Where the Pressure Cooker is used (the only Safe Continuous Method).	
			5 lb.	10 lb.
	Minutes.	Minutes.	Minutes.	Minutes.
Asparagus	5	180 (*not advised*)	60	40
Beans	5	180 (*not advised*)	60	30
Beets	6-10	120-180	60	40
Carrots	6-10	120-180	60	40
Cauliflower	5	180	60	40
Corn	5-10	180 (*not advised*)	90	40
Parsnips	6-10	180	60	40
Peas	5	180 (*not advised*)	60	30
Pumpkin	....	200 (*not advised*)	60	40
Spinach and other greens	5	200 (*not advised*)	60	40
Squash	....	200 (*not advised*)	60	40
Succotash	....	180	60	40
Meat	....	300	180	60

STORING CANNED GOODS.

(The Cornell Reading Course for the Farm Home.)

Canned food should be set aside for two or three days before storing, and then as a means of special precaution it should be tested as follows: Loosen the clamp and grasp the jar by the edge of the glass top. If the can leaks or if decomposition has set in, the top will come off, as is shown in Fig. 9. If the top stays on, tighten the clamp again and the food is ready for storage. If the top comes off, reject that can of food.

Canned food and vegetables should be stored in a dark place, as light destroys the colour, leaving the food unattractive in appearance. If the jar and its contents have been absolutely sterilized and the jar is entirely air-tight, the food will not spoil if held in a warm place. If spoiling does occur, it will be due to one of the following causes: (1) Some flaw in the can which makes it a slow leaker; (2) the presence of some microscopic organisms that have survived the cooking

Fig. 9. Manner of testing a jar.

process in spite of all care; (3) a drying-out of the rubbers and hence a breaking of the seal.

In some factories where foods are canned in glass jars racks are made for holding the jars upside down in an inclined position, thus keeping the liquid constantly in the top of the can and preventing the rapid drying of the rubber.

FERMENTATION AND SALTING.

The use of brine in preparing vegetables for winter use is much to be commended to the household. The fermentation method is in use in Europe, and is beginning to be better known in this country as a means of making sauerkraut and other food products.

No cooking is required by this process. Salt brine is the one requirement. The product may be kept in any container which is not made of metal and is water-tight. The vital factor in preserving the material is the lactic acid which develops in fermentation. An important feature is that vegetables thus prepared may be served as they are or they may be freshened by soaking in clear water and cooked as fresh vegetables.

Sauerkraut.

The outside leaves of the cabbage should be removed; then cut crosswise several times and shredded very fine with the rest of the cabbage. Immediately pack into a barrel, keg, or tub which is perfectly clean, or into an earthenware crock holding 4 or 5 gallons. The smaller containers are recommended for household use. While packing distribute salt as uniformly as possible, using 1 lb. salt to 40 lb. cabbage. Sprinkle a little salt in the container and put in a layer of 3 or 4 inches of shredded cabbage, and then pack down with a wooden utensil like a potato-masher. Repeat with salt, cabbage, and packing until the container is full or the shredded cabbage is all used. Press the cabbage down as tightly as possible and apply a cloth and then a glazed plate or a board cover. If using a wooden cover, select wood free from pitch. On top of this cover place stones or other weights (use granite and avoid the use of limestone or sandstone). These weights will serve to keep the brine above the cover.

Allow fermentation to proceed for ten days or two weeks if the room is warm. In a cellar or other cool place three to five weeks may be required. Skim off the film which forms when fermentation starts, and repeat this daily if necessary to keep this film from becoming scum. When bubbles cease to arise, if container is tapped, the fermentation is complete. If there is scum it should be removed. Pour melted paraffin over the brine until it forms a layer from ¼ to ½ inch thick to prevent the formation of the scum which occurs if the weather is warm or the storage-place is not well cooled. This is not necessary unless the kraut is to be kept a long time. The kraut may be used as soon as the bubbles cease to rise. If scum forms and remains the kraut will spoil. Remove scum, wash cloth cover and weights, pour off old brine, and add new. To avoid this extra trouble it is wise to can kraut as soon as bubbles cease to rise and fermentation is complete. To can, fill jars, adjust rubbers, and seal. Sterilize 120 minutes in hot-water pack or 60 minutes in steam-pressure outfit at 5 to 10 lb. pressure.

SALTING WITHOUT FERMENTATION.

Cabbage, string beans, and greens are preserved with salt. The amount of salt used will be one-quarter of the weight of the vegetables. Kegs or crocks make the best containers. Put a layer of vegetables about an inch thick on the bottom of the container. Cover this with salt. Continue making alternate layers of salt and vegetables until the container is almost filled. The salt should be evenly distributed, so that it will not be necessary to use more salt than the quantity required in proportion to the vegetables used. Cover the surface with a cloth and a board or a glazed plate. Press a weight on these and set aside in a cold place. If sufficient liquor to cover the vegetables has not been extracted by the next day, pour in enough strong brine (1 lb. salt to 2 quarts water) to cover surface around the cover. The top layer of vegetables should be kept under the brine to prevent moulding. There will be some bubbles at first. As soon as this stops set the container where it will not be disturbed until ready for use. Seal by pouring very hot paraffin on the surface.

HOME DRYING.

Home vegetable and fruit drying have been little practised for a generation or more, but during the past few years, with the high price of glass, tin, rubber, and fuel, many have found this method of preservation desirable. In the fruit districts of British Columbia, prunes, apricots, cherries, plums, peaches, and corn have been home-dried with splendid success.

The method is simple and practically all vegetables and fruits may be dried. The cost is slight, for in every home the necessary outfit in its simplest form is already at hand. Effective drying may be done on plates or dishes placed in the oven with the oven door partially open. It may be done on the back of the kitchen stove, with these same utensils, while the oven is being used for baking. It may also be done on sheets of paper or lengths of muslin spread in the sun and protected from insects and dust. A sheet of tin laid over a dripping-pan containing a small amount of hot water makes a good substitute for a certain type of commercial drier. The pan of water is kept over a slow heat sufficient to keep the water hot.

Barrel-hoops or frames made of laths may be covered with galvanized-iron netting or with cheese-cloth and suspended above the stove by a rope with a pulley arrangement, which makes it easy to adjust the trays at the proper height. Some housekeepers use window-screens on bricks as supports. Proper ventilation that allows for a free circulation of dry air is more important than heat in drying foods. For example, an electric fan placed before a dryer may accomplish excellent results without the aid of heat.

DRYING FRUITS.

All fruits that are to be dried should be well ripened, but not overripe. Fruits that are dried with the skins on should be dipped quickly by the means of a wire basket or a piece of cheese-cloth into a boiling solution of lye made in proportion of ½ lb. concentrated lye to 8 gallons water. They should then be rinsed two or three times in clear water. The lye perforates the skin and thus facilitates evaporation. Moreover, it destroys micro-organisms that might cause spoilage.

Most fruits are improved by being dipped into a thin syrup before being dried. If the fruit is to be used in puddings, cakes, breads, breakfast cereals, or as a confection, it may be sprinkled with sugar before being dried; if it is to be cooked for sauce, little or no sugar should be added.

Metal trays for drying should be covered with cheese-cloth to prevent acid action. Wrapping-paper may be used on trays in an oven.

Juicy fruits require more ventilation in drying than do such fruits as apples.

When fruit is sufficiently dry, it should be impossible to press water out of the freshly cut ends of the pieces. The natural grain of the fruit should not be apparent on cut surfaces. The fruit should be leathery or pliable, and not so dry that it will snap or crackle. In general, the drier the fruit the less chance there is for spoilage; but sweet fruits can safely contain more moisture than those with a low sugar content. Fruit should be cooled quickly after being dried in order to prevent a shrivelled and unattractive appearance.

DRYING VEGETABLES.

Equally as great care should be given to the selection and preparation of vegetables for drying as for canning. Good results depend largely on the use of vegetables that are absolutely fresh, young, tender, and perfectly clean. All vegetables should be washed and cleaned thoroughly before being dried. If steel knives are used for paring and cutting the vegetables, they should be kept clean and dry to prevent discoloration.

After being cleaned and prepared, the vegetables should be blanched as for canning, but not dipped in cold water. This removes a strong odour and flavour from certain kinds of vegetables and loosens the fibre, which allows the moisture in the vegetables to evaporate more quickly and uniformly. Moreover, it helps to retain the natural flavours and colour. After being blanched for the required number of minutes, the vegetables should be well drained and placed between two towels or exposed to the sun and air for a short time to remove the surface moisture.

The temperature at which most vegetables should begin drying after the surface moisture is removed is 110° Fahr.; and this should be gradually increased to 150° Fahr., which makes it possible in most cases to complete the drying in 2 or 3 hours.

STORAGE OF DRIED FOODS.

Dried fruits should always be stored in moist-proof containers and in a dry place free from dust and flies. The best container is a tin box, bucket, or can fitted with a perfectly tight cover. A glass jar with a tight seal is a good container for dried fruits. Paraffin-coated paper containers of various sizes can be found on the market.

When vegetables are first taken from the dryer, if completely dried, they are very brittle. They are more easily handled and are in better condition for storing if allowed to stand for from 1 to 3 hours to absorb moisture to make them pliable before they are stored. If they are allowed to stand for several days, they should be heated to 160° Fahr. to destroy any insect-eggs that may be on them, care being taken not to heat the vegetables higher than 160 degrees. Dried fruits should be examined occasionally in case of some infestation. Upon the first appearance of insects, the fruit should be spread in thin layers in the sun until the insects disappear; it should then be heated to a temperature of 160°. Fahr. and re-stored carefully.

Dried Corn.

Corn that is just right to use on the table should be selected for drying. This will be somewhere between the "milk" stage, in which the juice spurts out when the kernel is pressed open with the thumb-nail, and the "dough" stage, in which the contents of the kernel may be pressed out in a solid, soggy mass. Corn should be dried as soon as possible after it is gathered, and it should be dried as quickly as possible in order to retain the flavour. Left-overs of cooked corn on the cob may be mixed with sugar and salt in the proportions suggested and dried in the oven or over the stove.

METHOD 1.

5 quarts corn (measure after it has been cut from the cob).   ½ cup sugar.   ¼ ,, salt.

Strip off the husks and the silk and cut the kernels from the cob. Do not cut the kernels too close, but press out all the milk with the back of the knife. Mix the corn, sugar, and salt, place the mixture in a pan over a vessel of boiling water, and stir it frequently until all the milk has been absorbed. Spread the corn on plates and dry it in a slow oven, stirring it frequently to prevent scorching. It should be possible to complete the process in half a day, but a longer period may be necessary.

METHOD 2.

Remove the husks and the silk. Place the ears in boiling water for 5 minutes. With a sharp knife cut off the kernels and scrape the cob. Place the corn in thin layers on platters. Dry it in a slow oven or some other warm place, stirring it frequently.

Dried corn may be prepared for the table by any of the following methods:—

(1.) Soak the corn overnight in water. Heat it very slowly for 2 hours or until it is soft. Add milk, butter, and seasonings.

(2.) Put the corn in a double boiler, add to it about 1½ times its volume of cold water, and set it where it will very gradually heat to the simmering-point. From 2 to 3 hours will be required for proper cooking. To save fuel, soak the corn for 3 hours in the required amount of water, and then simmer it until it is tender. Avoid vigorous boiling. The corn approximately doubles its bulk during the preparation.

Dried String Beans or Wax Beans.

Wash the beans and carefully remove the strings from string beans. The very young and tender beans can be dried whole. Those that are full-grown should be cut in ¼- to 1-inch lengths with a sharp knife. They are then put into a bag of cheese-cloth or a wire basket and blanched in boiling water for from 6 to 10 minutes, depending on the maturity of the bean. One-half teaspoon soda may be added to each gallon of boiling water to help set the green colour. Remove the surface moisture according to the directions already given. Young string beans dry in 2 hours, more matured beans in 3 hours.

Spinach, Dried Herbs, and Seasonings.

(1.) Celery-tops, parsley, mint, sage, and herbs of all kinds need not be blanched, but they should be washed well, sliced and cut, and dried in the sun or in a drier. These are good for flavouring soups, purees, gravies, omelets, and the like.

(2.) Spinach and beet-tops may be steamed for 2 minutes before drying.

Dried Tomato Paste.

Tomatoes may be dried to a paste and used for soups, sauces, scalloped dishes, and the like. One teaspoon of the paste will make one dish of soup. The following method may be used: Blanch and skin the tomatoes. Slice and place in kettle to boil, adding no water. Boil the tomatoes until they are tender, rub them through a sieve, and boil down the pulp over direct heat until it is so thick it is difficult to cook without being stirred continually. Then place it over hot water or in a slow oven where there will be no danger of scorching; then put it where the moisture will evaporate until the pulp is stiff enough to hold its shape when lifted with a spoon.

It may then be placed in hot sterilized jars and sealed; or it may be spread on plates or pans in thin sheets and dried thoroughly in a very slow oven from 130 to 140° Fahr. until it can be cut in squares or rounds. It should then be stored carefully in moisture-proof containers.

Dried Cherries.

Dried cherries makes a most desirable substitute for raisins. They may be prepared in the following way: Wash and remove the surface moisture. Spread them, seeded or unseeded, in thin layers on trays. If the cherries are seeded the juice that they loose may be canned. If the skins of the cherries are tough and hard to dry they should be treated with a lye solution, which has been described. Dry them from 3 to 4 hours at 110 to 150° Fahr. Raise the temperature gradually.

Dried Plums.

Small thin-fleshed varieties of plums are not suitable for drying. Select medium-ripe plums, cover them with boiling water, cover the vessel, and let steam for 20 minutes. Drain them, remove the surface moisture, and dry them from 4 to 6 hours, gradually raising the temperature from 110 to 150° Fahr.

Dried Prunes.

The skin of prunes is usually thick and tough and so prevents the fruit from drying quickly. For this reason they are usually treated with a lye solution, as has been described, and then rinsed several times in clear water. They are dried in the same way as plums. If dried prunes are properly cooked they make one of the most delicious and pleasing fruit dishes. They should be soaked from 1 to 2 days. If they are soaked in freshly made tea their flavour is much improved and intensified. There should be liquid enough to cover them and they should be cooked slowly in the liquid in which they are soaked. The cooking should not take long when they are thoroughly soaked. When cooked add sugar enough to sweeten and a small amount of ground cinnamon if desired, or grated orange-rind. The pits may be cracked and the meats added to the fruit if desired.

Dried Peaches.

Peaches are usually dried unpeeled. Cut them in halves and remove the pits. Leave the fruit on trays with the pit side up, and dry them at 110 to 150° Fahr., raising the temperature gradually. Dry them for 4 to 6 hours, and longer if necessary.

Candied Fruit Peel.

The candied peel of oranges, grapefruit, and other citrous fruit make a sweet which is economical, and it utilizes material which otherwise might be thrown away. The skin may be kept in good condition for a long time in salt water, which makes it possible to wait until a large supply is on hand before candying them. The salt water takes out some of the bitter taste. The skins should be washed in clear water after removing from the salt water, boiled until tender, cut into small pieces, and then boiled in a thick sugar syrup until they are transparent. They should then be lifted from the syrup and allowed to cool in such a way that all superfluous syrup will run off. They should then be rolled in pulverized or fine granulated sugar.

VICTORIA, B.C.:
Printed by WILLIAM H. CULLIN, Printer to the King's Most Excellent Majesty.
1919.

Printed in the United States
122124LV00002B/148-150/A